Invention Algorithm (TRIZ) (A Workbook)

Creativity Method for Generating Breakthrough Inventions

Ted Rygas, Ph.D., P.Eng.

© 2017, Ted Rygas

3

Table of Contents

Table of Contents	4
Introduction	5
A. Important Concepts Needed for Effective use of the TRIZ Method	10
B. Formal Style in this Book	12
....	13
....	13
CHAPTER 1. Consider a Long Range Plan	13
CHAPTER 1. Consider a Long Range Plan	14
CHAPTER 2. Analyze "Present Technology Stage" (S-Curve)	16
CHAPTER 3. Analyze "Evolutionary Stage of Parameters"	17
CHAPTER 4. Analyze Physical Effects Relevant to the Invention	19
CHAPTER 5. MAIN PROCEDURE OF INVENTION ALGORITHM (FIVE STEPS)	21
Step 1. IDENTIFY PROBLEM TO BE SOLVED	21
Step 2. ANALYZE "ENVIRONMENT" OF SOLUTION	22
Step 3. ANALYTICAL PHASE	22
Step 4. PROBLEM SOLVING PHASE	23
Step 5. SYNTHETIC PHASE	24
CHAPTER 6. FORTY PRINCIPLES OF INVENTIONS	25
END OF THE ALGORITHM PART	38
....	38
....	38
APPENDIX A. LIST OF COMMON CONTRADICTIONS	39
APPENDIX B. TABLE OF PHYSICAL EFFECTS	41
APPENDIX C. LIST OF EFFECTS, ACCORDING TO THEIR APPLICATINS	43
APPENDIX D. INTERNET POINTERS AND RESOURCES FOR INVENTORS	60
APPENDIX E. SOFTWARE FOR INVENTORS	65
A. BOOKS FOR INVENTORS	66
B. GENERAL BIBLIOGRAPHY	66
APPENDIX F. FREQUENTLY ASKED QUESTIONS ABOUT TRIZ	68

Introduction

This Workbook will enable the users to create high-level creative solutions in solving technical problems. While the Invention Algorithm method is mostly applied to find solutions to engineering problems, it has been also successfully applied to such diverse fields as marketing or software development. The simple reason for the general applicability of the method is that it combines essentially all known approaches to getting creative solutions. The author of the method, Genrich Altshuller, extracted the principles of the Invention Algorithm from analyzing thousands of highly creative patents. He was interested in the patented solutions and he found, that there were only 40 generalized principles for solving technical problems. Additional work by other people researching creativity principles confirmed his finding: all creative technical solutions can be reduced to following one or more of those generalized principles. An abbreviation of the original Russian name of the Invention Algorithm method , TRIZ, became world-wide known name of this method.

Invention Algorithm (TRIZ) is being used by defense contractors, commercial companies and individual inventors.

Some companies, such as the South Korean "Samsung", achieved a spectacular commercial success as a result of implementing the TRIZ method to all of its operations. The South Korean "economical miracle" was in large part a result of embracing the Invention Algorithm method by that country. The 40 Universal Principles of Creativity are now finding applications not only in engineering, but also in business, social interactions, military weapon development, combat tactics and war strategies

The purpose of this Workbook is to make it easy for the user to use the full power of the Invention Algorithm (TRIZ) through an "on-the-job-self-training", while solving his technical tasks and developing a needed solution. In my professional experience, the self-training while solving your specific technical tasks is much more effective than taking a course and having the method presented by a training person. This Workbook will guide the user, step-by-step, to find creative and new solutions, often quite simple but often surprisingly ingenious, which are often overlooked in the industrial, manufacturing and even in the research environments.

The Invention Algorithm is much more than just a set of steps to follow until a breakthrough solution is achieved. Repeated reviewing of the creativity principles, as needed in solving the tasks, creates a skill of applying the principles to multiple areas of life, where a creative approach is needed. TRIZ also makes its user more open-minded and more analytical, as the problem solving enhances their skills of generalization and transferring the principles into real-world solutions. The power of the TRIZ method is also noticeable in removing mental blocks, preventing people from finding the creative solution. This "creativity mental block" is similar to the "writers' block", experienced when a person needs to write an essay or a technical report. The invention Algorithm breaks the "creativity mental block" by essentially forcing the user to see and describe an Ideal Final Solution and compare that Ideal Final Solution with the initial situation. Finding the creative solutions is then facilitated, as comparing the two situations leads to finding what needs to be done to have the problem solved. Solving the problem, the TRIZ way, with the Ideal Final Solution accepted, is then like building a bridge starting the construction from both sides of the river at the same time. The standard method, without the TRIZ method, can be compared to building the bridge starting from one side of the river, with the other side not assessed and covered in fog.

The skill gain due to acquiring of the TRIZ skills was well summarized by the designer of the Apple PowerBook,

David Levy, who said: "TRIZ is really how to be creative and to observe the world and solve problems".

This Workbook has been designed in a user-friendly checklist format, so the person using the Workbook will always know what is going to be his next step. The document is ready-to-annotate and contains examples guiding the user in his work through the easy to follow steps. The "consumable" checklists, brief reference cards, Internet links and other reference information are downloadable from the author's website on the internet (see the "Appendix - Annotated Bibliography"). The users are advised to keep the processed checklists with annotations in a binder. Keeping open a checklist, which was previously processed, will maintain the work organized and easy to restart. An additional benefit will be that the ser will be accumulating his past experience in an easy to find way.

Repeated use of the methodology results in embedding the creative thinking in the subconscious mind, and the user will notice that his/her thinking is using "creative principles" which enables the user to do "thinking in terms of the 40 principles". After several applications of the TRIZ, the user will be able to notice the opportunities for the application of the inventive principles in the projects on which he is working.

This Workbook was developed specifically to provide the users with an alternative to the standard training sessions, which typically provide the participants with bulky and difficult to follow training materials. Based on numerous cases observed by this author, the materials from those training courses typically end-up in some archive boxes. As a result, the traditional courses result in companies spending thousands of dollars on training, with providing very little or even zero help to engineers or scientists in their daily work.

The checklist format of this Workbook was designed in such a way, that the user will not skip any important part of the inventive work. Checklists are not only for the beginners; they are equally valuable for the experts in their fields (Gawande, 2011). Currently, the checklists are in use by the NASA (before lunching satellites into orbit), pilots, surgeons, skyscraper builders and other professionals. In his book, Gawande concludes that the checklists provide much needed assurance that all the critical issues are taken care of.

Checklists are also an ideal medium for accumulation and application of practical knowledge. This Workbook actually condenses the essential knowledge within the Invention Algorithm into an easy to follow checklist format.

In my professional experience, the most effective training occurs when a person works on his own tasks while equipped with an appropriate guidance. The checklist provided in this Workbook brings this essential guidance to the scientists and engineers working on their projects. The checklist format of the Workbook provides the reader with the advantage of saving his time. The consumable part of this Workbook is available for a free download from the author's website: https://sites.google.com/site/algorithmtriz .

A. Important Concepts Needed for Effective use of the TRIZ Method

There are only **two specific concepts** which need to be understood to effectively use this Workbook. These concepts are: the **"Technical Contradiction"** and the **"Ideal Final Result"**. These concepts will be illustrated using the inventive transition from a piston-type steam engine to a steam turbine.

The Technical Contradiction is an obstacle, which shows up if we try to increase certain desirable parameter. In the case of a piston-type steam engine, we want to increase the power output from that engine. The first approach is obvious and is not inventive: we just create a larger steam engine. We get more power, but the weight of the machine has increased significantly. This undesired increase in weight with increasing power represents a Technical Contradiction. In the classical language of TRIZ we have a <u>Contradiction between the parameter of power and the parameter of weight.</u>

Now, we will consider what we really want from the power generating machine: we want a shaft which will rotate and which will provide a torque to rotate the wheels of a locomotive or which will keep turning the propeller of a ship. Our **Ideal Final Solution** (IFS) is a shaft, rotating and providing useful power. At this point, we need to imagine that we have a shaft, perhaps supported by some bearings. The bearings are needed to keep the shaft in place. <u>Nothing more</u> should be assigned to the achieved Ideal Final Solution at this stage.

Looking at the evolutionary trends of inventions, we now see that the direction of the evolution is to move from linear motions to rotations. Therefore, we should be looking for a solution preferably avoiding the pistons. Now, reviewing the standard 40 Principles of Inventions, we see that the following principles should be considered in making a breakthrough invention:

1. #14 The Principle of Spherical Shape
2. #20 The Principle of Uninterrupted Useful Effect.
3. #29 Use Pneumatic or Hydraulic principles

4.   #37 The Principle of Heat Expansion

At this point, the inventor is guided to the solution where the steam is directly blowing on the blades and moving the shaft. The blades, although not of a spherical shape, are making the circular motions, at least partly resembling the sphere. Typically, we only expect some hints from the 40 Principles and not a perfect match. The inventor still has to be creative in adapting the Principles. The principle of uninterrupted useful effect will be achieved, as the rotations of the turbine are continuous. The pneumatic principle, in this case interpreted as expansion of steam, is also used. The last principle, needs to be interpreted in general terms: heating the steam causes steam expansion, and if the volume is restricted, as in a boiler, the effect is a pressure build-up. The steam expands as it flows toward the blades of the turbine.

B. Formal Style in this Book

The style used in this book is simplified with respect to using the gender designations. In this book, the convention is that the words "he" or ""his" are used instead of cumbersome words with slashes, such as "he/she" or "his/her". This should not be interpreted as a lower regard for one of the genders.

OUTLINE OF THE INVENTION ALGORITHM – MAIN ACTIONS

CHAPTER 1. Consider a Long Range Plan	14
CHAPTER 2. Analyze "Present Technology Stage" (S-Curve)	17
CHAPTER 3. Analyze "Evolutionary Stage of Parameters"	18
CHAPTER 4. Analyze Physical Effects Relevant to the Invention	20
CHAPTER 5. MAIN PROCEDURE OF INVENTION ALGORITHM (FIVE STEPS)	22
Step 1. IDENTIFY PROBLEM TO BE SOLVED	22
Step 2. ANALYZE "ENVIRONMENT" OF SOLUTION	23
Step 3. ANALYTICAL PHASE	23
Step 4. PROBLEM SOLVING PHASE	24
Step 5. SYNTHETIC PHASE	25
CHAPTER 6. FORTY PRINCIPLES OF INVENTIONS	26
END OF THE ALGORITHM PART	39

CHAPTER 1. Consider a Long Range Plan

Specify range and parameters of the planned invention

1.1. Specify the name of the product or solution to be invented
....
....

1.2. Specify the required parameters:
- size: ....
- weight: ....
- speed: ....
- reliability: ....
- maintenance: ....

1.3. Declare the range of the project:
   - magnitude of change:
- small: ....
- large: ....
  - project manpower: ....
  - costs/funds: ....

The magnitude of change, according to TRIZ, depends on resolving contradictions, applications of various fields of technology and use of rare physical effects:

Table 1. Invention levels defined according to TRIZ

Level of invention	Contradiction resolved?	Fields of technology	Use of rare physical effects
1	Unresolved	Single	Not used
2	Resolved with small change	Single	Not used
3	Resolved with a major change.	Several	Not used
4	Resolved; the object is totally changed	Several	Use of rare physical effects
5	Resolved using a new fundamental effect	Several	New physical effect discovered (for example discovery of a laser).

CHAPTER 2. Analyze "Present Technology Stage" (S-Curve)

Stage 1. Infancy - all parts are working without problems, at least one element is controllable.

....
....

Stage 2. Optimization and rapid growth of technical parameters.

....

Stage 3. Maturity and dynamization: elements of the system are changing to adapt to
    changing conditions.

....
....

Stage 4. Maturity and self-monitoring: automatic control and feedback are used.

....
....

Stage 5. Stagnation. The device/invention tends to undergo a transition to a higher level system:
(1) Macro to micro: mechanical parts are replaced with micro particles, molecules or liquids);

....
....

(2) Systems are joined to create a super-system (e.g. catamaran, computer networks).

....
....

CHAPTER 3. Analyze "Evolutionary Stage of Parameters"

3.1. Complexity:
- Mono-system ....
- Poly-system ....
- Poly-system with self control ....
- Poly-system with multiple controls ....

3.2. Functions performed by the device/invention:
- Single-function ....
- Single-function with adjusting of properties ....
- Multi-functional system ....
- Multi-functional system with self control ....
- Multi-functional system or network with multiple controls ....

3.3. Mobility:
- Immobile part ....
- Part with a joint ....
- Part with many joints ....
- Elastic system ....
- System using force field ....

3.4. Physical state:
- Solid
  ....
- Liquid
  ....
- Gas
  ....
  Force field
  ....

3.5. Composition:
- Monolithic part
  ....
- Part with a void
  ....
  Part with many voids
  ....
- Miniature voids
  ....
- Dynamic voids
  ....
  ....

3.6. Shape: • Flat .... • Bent .... • Curved .... • Changing shape ....

3.7. Actions:
- Continuous action
  ....
- Vibrations
  ....
- Resonance and standing waves
  ....
- Dynamically controlled waves
  ....

CHAPTER 4. Analyze Physical Effects Relevant to the Invention

In the TRIZ method, the physical effects are <u>listed according to their applications and functions</u>; for example, sub-listings may provide effects suitable for checking properties of surfaces, detection of materials, control of movement etc.

PHYSICAL EFFECTS

* Website support for this Workbook:
https://sites.google.com/site/algorithmtriz
* Physical effects with links to descriptions:
http://en.wikipedia.org/wiki/List_of_effects   (Very Good)
* Electrical Effects and descriptions:
http://en.wikipedia.org/wiki/Electrical_phenomenon   (Very Good)
* Optical Effects and descriptions:
http://en.wikipedia.org/wiki/Optical_phenomenon   (Very Good)
* Surface analysis methods:
http://en.wikipedia.org/wiki/List_of_surface_analysis_methods   (Very Good)
* Thermal analysis methods:
http://en.wikipedia.org/wiki/List_of_thermal_analysis_methods   (Very Good)
* Inventors' resources, compiled by Ronald J. Riley:
http://www.rjriley.com/site-index/

A large list of physical effects is attached to this Workbook (see APPENDIX B" - Physical Effects). An updated list of physical effects, grouped according to their application, is also available on the Workbook Support Webpage:

https://sites.google.com/site/algorithmtriz

The field of physical effects is still growing with new physical effects being discovered.

Another approach to finding the relevant physical effects is to use TRIZ software (see section 5, below).

CHAPTER 5. MAIN PROCEDURE OF INVENTION ALGORITHM (FIVE STEPS)

A session following this practical approach should last for about 60 minutes. It is recommended that the inventor proceeds with his first session on his own. A group session of 3 - 5 participants may follow. Follow the FIVE STEPS of the algorithm to get full benefit of this method. The FIVE STEPS have references to other elements of the method. The most simple and the most effective way of using this technique (recommended for the first year of using "TRIZ") is to review all 40 PRINCIPLES in each case. Make notes as you proceed.

Step 1. IDENTIFY PROBLEM TO BE SOLVED

1.1 What is the expected final solution (ideal machine, ideal state, ideal properties etc.; for example: part suspended in the air and not falling down).

1.2 Define an alternative ("go around") solution.

1.3 Find which solution, straight or alternative, will give better results. Compare both solutions with development trends. Make a selection (straight or substitute solution).

1.4 Define detailed technical parameters (yield, dimensions, etc.)

1.5 Find out if higher than "minimum required" parameters are achievable (check for future capabilities).

1.6 Find out if the selected solution is not too complex.

NOTES:

....
....
....
....

Step 2. ANALYZE "ENVIRONMENT" OF SOLUTION

2.1 Check patent literature for solutions to similar and opposite problems.

2.2 Analyze how conditions of the solution will change with full range (0 to infinity) of:
  - dimensions; - time; -speed; -velocity; - funds; -cost.

2.3 Rank above listed elements according to our ability to change them.

2.4 State the problem in plain words (no technical terms). Are the initial assumptions changing?

Step 3. ANALYTICAL PHASE

3.1. Define what is the ideal solution. Make two drawings showing all elements BEFORE and AFTER.

3.2. List what prevents you from getting the ideal solution.

3.3. Find out what causes that the ideal solution is not achieved (STATEMENT OF CONTRADICTION).

3.4. State under which conditions the contradicting conditions will disappear (compensated or removed).

3.5. Find out a method or a device to remove the contradiction (solid/liquid/gas ?; changes in operation ?).

Before   After

_____

Step 4. PROBLEM SOLVING PHASE

4.1 Check the table of "Detailed Forty Principles" (main strength of the method; author's recommendation is to review the complete list, starting from the top principles).

4.2 Find out if the environment of the object can be changed/modified.

4.3 Find out if objects interacting with the given object can be changed.

4.4 Find out if changing the time factor will solve the problem (faster, slower, in-advance action, after work action; impulse/batch/continuous action).

4.5 Find out how similar problems were solved in the animal world (extinct/present species; what are the trends). (This is a powerful approach, but may require high-level external consultation).

Step 5. SYNTHETIC PHASE

5.1 How other elements of the object will be affected by the proposed solution ?

5.2 List changes required in the cooperating objects.

5.3 Could the modified object be used in other applications?

5.4 Apply the new solution to other technical problems.

CHAPTER 6. FORTY PRINCIPLES OF INVENTIONS

☐ **1.0 The Principle of Segmentation (and Reverse: Consolidation).**

1.1 Divide the object into independent parts. Divide it further into a powder, a liquid or a gas.

1.2 Make the object divisible or easily dismantled.

1.3 Divide the object into a group of joined identical objects, such as a bundle of steel pipes rather than individual holes drilled in a solid block (metal heat exchanger example). Or, try the reverse (graphite block heat exchanger).

☐ **2.0 The "Take-Away" Principle.**

2.1 Take away the problem-creating part of the object.

2.2 Remove the useful part of the object and allow the rest to remain.

☐ **3.0 The Principle of Local Quality (weak principle).**

3.1 If the structure is homogeneous, make it non homogeneous.

3.2 A part of the object carries out the required function.

3.3 Each part of the object is operated in the environment which is best suited for its operation.

☐ **4.0 The Principle of Asymmetry.**

4.1 Use asymmetrical rather than symmetrical features.

4.2 If it is already asymmetrical then increase the degree of asymmetry.

☐ **5.0 The Principle of Merging Similar or Dissimilar Objects.**

5.1 Join homogeneous objects.

5.2 Join objects designed for continuous operation.

5.3 Perform operations in parallel.

5.4 Incorporate similar objects into the same system. For example, "bundle" together a large number of insects to measure their temperature using a regular thermometer. Connect two bikes to make a tandem. Connect two boats to make a catamaran. A bundle of tubes can be used as a pressure vessel for very high pressures.

5.5 Merge dissimilar objects which are normally found together, such as a pencil and eraser.

5.6 Merge dissimilar objects in such a way that a new feature appears. For example, a boat and a car or an airplane and a submarine. Occasionally, a transition to another dimension may help. (Production rate may be improved by using sequential (in-line) workstations).

☐ **6.0 The Principle of Universality (the "Swiss Knife" Principle).**   6.1 One object performs multiple functions, which eliminates needs for other parts.

☐ **7.0 The Principle of Nesting Dolls.**

7.1 One object is placed inside another object, which is in turn placed inside another object.

7.2 One object passes through a hole or cavity in another object.

7.3 Parts of one object are located inside another part.

☐ **8.0 The Principle of Counter-Weight.**

8.1 Compensate for the weight or an undesirable force on an object by combining this object with a counter-weight or counter-force which nullifies the undesirable force. Check if hydraulic or aerodynamic forces (for example servo-mechanisms) could be of use. Examples: counter-weights used on elevators; power steering and power breaks in cars.

☐ **9.0 The Principle of Preliminary Counter-Action**

(stronger than principle no. 10).

9.1 Before some action is performed on an object, a preliminary action is performed which directly counters the first action and enhancing the desired effects.

☐ **10.0 The Principle of Preliminary Action.**

10.1 If the required action can not be carried out, then carry it out in advance, either partially or fully.
 This may require use of a tool, or a substance, fully or partially placed in advance.
10.2 Arrange the objects in such a pattern, that some other action can be carried out in the most efficient way.

☐ **11.0 The Principle of "Previously Placed Pillow".**

11.1 Compensate for the low reliability of a component by using a supplementary back-up component which will perform the function of the first part, if the first part fails.

☐ **12.0 The Principle of The Same Potential.**

12.1 Perform the operation in such a way, that no lifting or lowering is required.

☐ **13.0 The Principle of "The Other Way Around" (strong principle).**

13.1 Instead of the expected action required by the problem statement, do the reverse.

13.2 Use the opposite feature or property. (Use white on black rather than black on white. Push rather than pull. Cool rather than heat).

13.3 Switch the moveable part of the object to be immovable and the immovable part to be movable.

13.4 Turn the object upside-down, or inside out, or reverse it.

☐ **14.0 The Principle of Spherical Shape.**

14.1 Switch from flat design to spherical design.

14.2 Use rollers, bearing, or spirals.

14.3 Switch from linear to rotating movement.

14.4 Use centrifugal force.

☐ **15.0 The Principle of "Dynamism".**

15.1 The characteristics of an object change so as to be the optimum at each stage. For example, a vertical start airplane with changing positions of engines.

15.2 The object is divided up in such a way that each part can be repositioned to perform the optimal function.

15.3 Parts of an object that are not interchangeable are redesigned into interchangeable.

☐ **16.0 The Principle of Partial or Excessive Action (strong principle).**

16.1 Apply somewhat more than is required, and then remove the excess.

16.2 Apply somewhat less than is required, then later add more.

16.3 If the activity cannot be done completely, then do it partially. For example, place a tube for a saw blade in the cast, before it hardens, to facilitate removal of the cast later.

☐ **17.0 The Principle of Moving to Another Dimension.**

17.1 An action that is performed in one dimension is changed to a higher or lower dimension.

17.2 An action that is performed in a plane is changed to one that is performed in 3 dimensions.

17.3 Use an assembly of many layers rather than one layer.

17.4 Change the orientation of the object.

☐ **18.0 The Use of Mechanical Vibrations.**

18.1 Set the object vibrating.

18.2 If the object is already vibrating then increase the amplitude of the oscillation.

18.3 Make use of the natural frequency of the object.

18.4 Use piezo vibrators rather than mechanical vibrators.

18.5 Make use of ultrasonic vibrations.

☐ **19.0 The Principle of Periodic Action** (weaker than principle 20).   19.1 Rather than a continuous operation, move to a periodic operation.

19.2 If the action is already periodic, then change its period.

19.3 Use the pauses between each period to perform some other useful purpose.

☐ **20.0 The Principle of Uninterrupted Useful Effect.**

20.1 Change a periodic operation to a continuous one.

20.2 Remove "dummy/idle" runs.

☐ **21.0 The Principle of Rushing Through (The "Skip" Principle)**

21.1 Perform harmful or dangerous operations in extremely short time or at a very high speed.

21.2 "Skip" through the dangerous region. For example rush through the resonant frequency when accelerating rotating equipment.

☐ **22.0 The "blessing in disguise" principle or "turning harm to good" (strong principle).**

22.1 Combine two or more harmful factors to create a good effect.

22.2 Use an effect rather than fight against it.

22.3 Add several harmful factors to create a useful one.

22.4 Increase a harmful effect until it is no longer harmful.

☐ **23.0 The Feedback Principle.**

23.1 Introduce feedback.

23.2 If there is already feedback, then change it or enhance it.

☐ **24.0 The "Go-Between" Principle. The Principle of an Intermediate Object**

24.1 Use an intermediate object to transmit the action.

24.2 Temporarily join two objects and remove the joint later.

☐ **25.0 The Self-Service Principle.**

25.1 The device services itself by performing auxiliary or repair functions.

25.2 Substance is replaced or refreshed automatically upon usage.

25.3 Use waste materials or energy.

☐ **26.0 The Copying Principle.**

26.1 Use a cheap copy of the object rather than the original which may be expensive, complex, delicate, unavailable, fragile, or inconvenient.

26.2 Replace an object or system of objects with an optical copy such as a photograph, photocopy, inverse copy, casting or mold.

26.3 If visible copies are available then switch to infra-red or ultraviolet copies.

☐ **27.0 Cheap Short Life Rather Than Expensive Long Life (use disposable objects).**

27.1 Replace an expensive object which has an expected long life with disposable objects.

☐ **28.0 Replace a Mechanical Pattern (strong principle).**

28.1 Replace a mechanical pattern with an optical, acoustical or "smell" pattern.

28.2 Use an electrical, magnetic or electro-magnetic field to replace a mechanical field.

28.3 Change from a field which is immovable to one which is movable.

28.4 Change from a field which is unchanging to a dynamic one.

28.5 Change from a field which is unstructured to a structured one.

28.6 Use a field-force in combination with ferromagnetic particles (very strong).

☐ **29.0 Use Pneumatic or Hydraulic principles, parts or mechanisms.** 29.1 Use gas or liquid parts of an object rather than solid parts.
29.2 Use components which are inflatable.

☐ **30.0 Use flexible membranes and fine membranes.**
30.1 Instead of normal constructions, use flexible or fine membranes.
30.2 Use soap bubbles, foam or films.

☐ **31.0 Use Porous Materials (or use voids/empty space).**
31.1 Make the object or part of the object porous.
31.2 Fill the holes in a porous substance in advance.

☐ **32.0 The Principle of Using Paint.**
32.1 Change the color of an object.
32.2 Change the color of its surroundings.
32.3 Change the transparency of an object or its surroundings.
32.4 Use colored additives to make something more visible.
32.5 Use luminescent traces.
32.6 Use soap bubbles or foam.

33.1 Objects interacting with a given object should be made of the same material or material with closely matching properties.

☐ **34.0 The Principle of Discarding and Regenerating Parts (strong principle).**

34.1 Parts used and no longer needed, should be discarded, dissolved, evaporated or changed in shape.

34.2 Parts which are used up should be regenerated as they are in use (for example a sharpening cycle in a cutting machine).

☐ **35.0 The Principle of Changing the Physical or Chemical State of an Object.**

35.1 Change from a solid to a liquid or from a liquid to a solid.

35.2 Change from a liquid to a gas or from a gas to a liquid.

35.3 Freeze rather than heat. Do the reverse chemical or physical state.

35.4 Change from one state to an intermediate state.

35.5 Change to a pseudo state such as an elastic.

35.6 Change to a bi-phase state such as boiling.

35.7 Dissolve or use phase transitions.
Examples are: Further freeze the ice to break it up.
Use dry ice to "sand blast".

Use ice slush jets rather than high pressure water jets to peel potatoes.

☐ **36.0 Using Phase Changes (strong).**

36.1 Use properties changing with phase transitions, such as volume, heat capacity, shape or absorption capacity.

☐ **37.0 The Application of Heat Expansion.**

37.1 The use of the expansion or contraction of a material with the application of heat or cold.

37.2 Make use of the different rates of expansion or contraction characterizing various materials. (Example: metal with thermal expansion coefficient equal zero - Invar).

☐ **38.0 Use Strong Acidifiers (or Materials with High Chemical Reactivity/Energy).**

38.1 Replace normal air with enriched air.
38.2 Replace enriched air with oxygen.
38.3 Ionize the air or oxygen.
38.4 Use ionized oxygen.

☐ **39.0 Use an Inert Environment.**

39.1 Carry out an action in an inert gas rather than air.

39.2 Carry out the operation in a vacuum.

☐ **40.0 Use Composite Materials.**

40.1 Switch from homogeneous material to composites.

SKETCHES – IDEAS

Principles to consider:

....

....

....

....

....

....

~~~~ END OF 40 PRINCIPLES - GO BACK TO THE FIVE STEPS (MAIN PROCEDURE) ~~~~

END OF THE ALGORITHM PART

....

....

APPENDIX A. LIST OF COMMON CONTRADICTIONS
(39 parameters)

Lists of TRIZ contradictions and the most often used resolving principles (selected from the "40 Principles of Inventions") are available at the following websites:
(1) http://creatingminds.org/tools/triz/triz_contradiction_1.htm (TRIZ contradiction matrix) and
(2) http://www.triz40.com/ - drop-down list of TRIZ contradiction matrix.

This author recommends reviewing of all 40 principles of creativity for every task rather than taking into consideration just 3 to 4 of them. This scanning through all the principles will develop the best understanding of the TRIZ method. Using the limited number of principles from the table of contradictions carries a risk that in your specific example an important principle may be missed.

The 39 contradiction parameters are listed below:

1. Weight of moving object
2. Weight of non-moving object
3. Length of moving object
4. Length of non-moving object
5. Area of moving object
6. Area of non-moving object
7. Volume of moving object
8. Volume of non-moving object
9. Speed
10. Force
11. Tension, pressure
12. Shape
13. Stability of object
14. Strength

15. Durability of moving object
16. Durability of non-moving object
17. Temperature
18. Brightness
19. Energy spent by moving object
20. Energy spent by non-moving object
21. Power
22. Waste of energy
23. Waste of substance
24. Loss of information
25. Waste of time
26. Amount of substance
27. Reliability
28. Accuracy of measurement
29. Accuracy of manufacturing
30. Harmful factors acting on object
31. Harmful side-effects
32. Manufacturability
33. Convenience of use
34. Repair-ability
35. Adaptability
36. Complexity of device
37. Complexity of control
38. Level of automation
39. Productivity

APPENDIX B. TABLE OF PHYSICAL EFFECTS

After finding an appropriate effect, go to Assessment of Solution ("Five Steps", Step 4).
This table is a collection of physical effects. It helps in locating of various, sometimes rarely used, physical effects, which could help in solving the contradiction. All possible effects relevant to the case should be studied and carefully considered.

More details on the Physical Effects are located at:

* Invention Algorithm (TRIZ) – supplement to this book (author's site):
    https://sites.google.com/site/algorithmtriz

* Physical effects with links to descriptions:
http://en.wikipedia.org/wiki/List_of_effects (Very Good)

* Electrical effects and descriptions:
http://en.wikipedia.org/wiki/Electrical_phenomenon (Very Good)

* Optical effects and descriptions:
http://en.wikipedia.org/wiki/Optical_phenomenon (Very Good)

* Surface analysis methods:
http://en.wikipedia.org/wiki/List_of_surface_analysis_methods (Very Good)

* Thermal analysis methods:
https://en.wikipedia.org/wiki/Thermal_analysis (Very Good)

* Physical effects for inventors: http://triz.it/eng/ebf/vrg08.htm (Good)

* Chemical effects for inventors:
http://triz.it/eng/ebf/vrg09.htm (Good)

* Geometrical effects for inventors:
http://triz.it/eng/ebf/vrg10.htm (Can be useful)

* Inventors' resources, compiled by Ronald J. Riley:
http://www.rjriley.com/site-index/ (Good)

GUIDING TABLE – CONTROL OF PARAMETERS

TEMPERATURE	POSITION and FLOW	POSITION CONTROL
Measuring temperature Lowering temperature Raising temperature Stabilizing temperature	Indication of position and location of object Measuring dimensions of objects Control of aerosol flows Changing the dimensions of objects Controlling location of objects	Control of movement Control of aerosol flows (dust, fog, smoke) Stabilization of position of object Action of forces. Control. Creation of high pressures Setting up interaction of mobile, (exchangeable), and immobile, (fixed), objects
SURFACE and VOLUME PROPERTIES	**ENERGY TRANSFER and DESTRUCTION**	**ELECTROMAGNETIC RADIATION / LIGHT**
Changes in friction Checking of state and properties of surfaces Measuring surface properties Inspection of state and properties in volume Changing the volume properties of an object Creating a given structure. Stabilization of structure of an object	Destruction of object Accumulation of mechanical and heat energy Transfer of energy Action of forces	Indications of electrical and magnetic fields Indications of radiation Generation of electromagnetic radiation Control of electromagnetic fields Controlling light, light modulation
MIXING, SEPARATING, CHEM. CONVER.	**MISCELLANEOUS**	
Forming mixtures Separation of mixtures Initiation and intensification of chemical changes	Geometrical effects	

APPENDIX C. LIST OF EFFECTS, ACCORDING TO THEIR APPLICATINS

Measuring temperature:

- Heat distribution and the change it causes in the object's internal frequency of vibration.
- Thermo-electrical phenomena (mostly thermocouples).
- Spectrum of radiation (including IR sensors).
- Changes in optical, electrical, magnetic properties of substances.
- Move through the Curie point.
- Hopkins effect (effect (an increase in magnetic susceptibility slightly below the Curie temperature. At the Curie point, the magnetic susceptibility disappears.
- Barkhausen effect (stepwise, small jumps in magnetic susceptibility during magnetization or demagnetization)

Lowering temperature:
- Phase transitions (condensation, freezing, crystallization).
- Joule-Thomson effect . This effect causes cooling of real gasses due to expansion through a throttle or a porous membrane. Under certain conditions this effect may result in heating of the fluid; this happens usually at very high pressures.
- Rankin Cycle.
- Magnetocaloric effect (magnetic refrigeration).
- Thermoelectric phenomena (thermocouple phenomena).

Raising temperature:
- Electromagnetic induction.
- Eddy currents.
- Surface effect.
- Dielectrical heating.
- Electronic heating using radio-frequency radiation.
- Electrical resistance heating. This can use the transformer effect.

- Absorption of various radiations by the substance.
- Thermo-electrical phenomena.

Stabilizing temperature:
- Phase transitions (condensation and evaporation under controlled pressure; the move through the Curie point).

Indication of position and location of object:
- Introduction of marker substances (security tracers etc.);
- Transforming the external fields (luminescent traces) or creating their own fields, such as ferromagnetic which are easily inspected (RF-ID).
- Reflection and emission of light.
- Photoelectric effect.
- Deformation.
- X-ray and radioactive radiation.
- Luminescence.
- Change in electrical and magnetic fields.
- Electrical discharges.
- Doppler effect.

Controlling and changing location of objects:
- Action of magnetic field on object or on ferro-magnet linked to the object.
- Action of electrical field on charged object.
- Transfer of pressure of liquids and gases.
- Mechanical oscillations.
- Centrifugal forces.
- Heat distribution.
- Light pressure.

Control of movement:
- Capillary action.
- Osmosis.
- Toms effect (drag reduction in fluid flow caused by addition of small amounts of polymer to that fluid) .
- Bernoulli effect (increase in the speed of a fluid causes drop of pressure).

- Wave movement.
- Centrifugal forces and gyroscopic effects (maintenance of the direction of the axis of rotation in spinning masses)
- Weissenberg effect (viscous liquid climbing up of a rotating rod inserted into that liquid).

Control of aerosol flows (dust, fog, smoke):
- Use of electrical charges.
- Electrical and magnetic fields.
- Light pressure.
- Air vibrations.
- Liquid drench.
- Use of demisters.

Forming mixtures:
- Mechanical agitation (mixers).
- Cavitations (including homogenizers and passing through an orifice).
- Diffusion.
- Electrical fields.
- Magnetic field combined with a ferromagnetic substance.
- Electrophoresis (suitable for separations as well).
- Solubilization with a third component.

Separation of mixtures:
- Electrical and magnetic separation.
- Changing apparent viscosity of the separator liquid under the effect of electrical and magnetic fields.
- Centrifugal forces (centrifuges, ultracentrifuges, cyclones, hydrocyclones).
- Sorption (selective; molecular sieves etc.).
- Diffusion.
- Osmosis (forced using pressure).
- Flotation
- Pressure swing distillation
- Extractive distillation
- Freezing-out and fractional crystallizations.

- Electrophoresis (bio-separations).

Stabilization of position of object:
- Electrical and magnetic fields.
- Fixing in liquids which harden in magnetic and electrical fields.
- Hygroscopic effect.
- Reactive movement.

Action of forces. Control. Creation of high pressures:
- Effect of a magnetic field via a ferromagnetic substance.
- Phase transitions.
- Heat distribution.
- Centrifugal forces.
- Changing the hydrostatic forces by changes in the apparent viscosity of magnetic
    or electrically conducting liquids in a magnetic field.
- Use of explosives.
- Electro-hydraulic effect.
- Optical hydraulic effect.
- Osmosis.

Changes in friction:
- Johnsen-Rahbeck effect.. Application of an electric field between a metallic and a semiconductor surface causes an increase in the friction coefficient).
- Action of radiation.
- Kragelsky phenomenon.
- Oscillation.
- Rheological shear-thinning.
- Rheopectic materials. Rheopectic effect is the increase of viscosity on exposure to shear; this property is rare.
- Drag reduction effect (high molecular weight polyethylene oxide, ppm concentration, in water). Applications: pumping of water, naval vessels; also kown as Toms effect.

Destruction of object:

- Electrical discharges.
- Electro-hydraulic effect.
- The hydraulic shock effect. This effect is also known as the "water hammer effect". It happens when a valve on the end of a pipeline suddenly closes. The closure causes a high energy pressure wave.
- Resonance.
- Ultrasonics.
- Cavitation.
- Induced radiation.
- Laser ablation.
- Explosive materials.
- Accelerated corrosion or dissolving.

Accumulation of mechanical and heat energy:
- Elastic deformations.
- Gyroscopic effect (kinetic energy of rotating mass).
- Phase transitions.

Transfer of energy:
- Deformations.
- Oscillations.
- Alexandrov Effect.
- Wave movement including electric shock waves.
- Radiation.
- Heat conductivity.
- Convection.
- Phenomenon of reflection of light, (light carriers).
- Induced radiation.
- Electromagnetic induction.
- Superconductivity.

Setting up interaction of mobile, and immobile, objects:
- Use of electromagnetic fields, (transition from "substance" to "field").
- Use of pneumatic forces (triggered air puff).

Measuring dimensions of objects:

- Measuring inherent frequency of oscillation.
- Applying and reading magnetic, optical or electrical markers.

Changing the dimensions of objects:
- Heat distribution.
- Deformation.
- Magnetostriction or electrostriction. This is a change in the shape of a three-dimensional object exposed to a magnetic or electric field. Non-conducting materials have this property to various degrees.
- Piezoelectric effect. Some materials (typically quartz) are creating strong static electrical charges on mechanical compression.

Checking of properties of surfaces:
- Electrical discharge.
- Reflection of light.
- Electronic emissions.
- Moire effect.
- Radiation.
- Methods of surface analysis:
  http://en.wikipedia.org/wiki/List_of_surface_analysis_methods

Measuring of surface properties:
- Friction.
- Absorption.
- Diffusion.
- Bauschinger effect. An example: cold working of steel increases the tensile yield strength and decreases the compressive yield strength.
- Electrical discharges.
- Mechanical and acoustic oscillations.
- Ultraviolet radiation.

Inspection of state and properties in volume:

- Introduction of "marker" substances transforming the external fields, (luminescent traces),
    or creating their own fields, (ferromagnetic), dependent on the state and properties
        of the substance under study.
- Changing the mean electrical resistance depending on the structure and properties of the object.
- Interaction with light.
- Electric and magnetic optical phenomena.
- Polarized light.
- X-ray and radioactive radiation.
- Electronic paramagnetic and nuclear magnetic resonance.
- Magnetic remnance effect.
- Move through the Curie point.
- Hopkins and Barkhausen effects. Changes in magnetic susceptibility with temperature.
- Measuring the inherent frequency of oscillation of an object.
- Ultrasonics.
- Moessbauer effect. This is a recoil-free emission or absorption of gamma photons by nuclei bound in a solid.
- The Hall effect. The deflective effect of an external magnetic fields on the of current flowing in a material.

Changing the volume properties of an object:
- Changing the properties of liquids, (apparent viscosity, fluidity), under the action
    of electrical and magnetic fields.
- Heat action.
- Phase transitions.
- Ionization under the effect of an electrical field.
- Ultraviolet, X-ray, radioactive radiation.
- Deformation.
- Diffusion.
- Electrical and magnetic fields.
- Bauschinger effect. An example: cold working of steel increases the tensile yield strength and decreases the compressive yield strength.

- Thermoelectrical, thermo-magnetic and magneto-optical effects.
- Cavitation.
- Photochromic effect. Color change of a material due to exposure of the material to the UV-light. Examples are spiropyranes and fulgides.
- Internal photo-effect. This is an increase in conductivity on exposure of a material to strong light.

Creating an interference structure:
- Interference waves.
- Standing waves.
- Moire effect.
- Magnetic waves.
- Phase transitions.
- Mechanical and acoustic oscillations.
- Cavitation.

Indications of electrical and magnetic fields:
- Osmosis.
- Electrical charging of bodies.
- Electrical discharges.
- Piezo- and magneto-electrical effects.
- Electrets.
- Electronic emissions.
- Electro-optical phenomena.
- Hopkins and Barkhausen effect.
- Hall effect.
- Nuclear magnetic resonance.
- Gyromagnetic and magnetic optical phenomena.

Indications of radiation:
- Optical acoustic effect.
- Heat distribution.
- Photoeffect.
- Luminescence.
- Photoplastic effect.

Generation of electromagnetic radiation:
- Josephson effect. A macroscopic quantum phenomenon, where current in one superconductor causes appearance of current in another, close but isolated superconductor. one superconduc
- Induced radiation.
- Tunneling effect.
- Luminescence.
- Hahn effect. An effect used in nuclear magnetic resonance spectroscopy.
- Cherenkov effect.
- Stokes and anti-Stokes effects. Stokes effect is the regular fluorescence effect with the light emitted being of a longer wavelength that the excitation light. Anti-Stokes effect is when the material emits fluorescent light of the shorter wavelength than the wavelength of the excitation light.

Control of electromagnetic fields:
- Screening.
- Changing state of environment, for instance, increasing or decreasing its electrical conductivity.
- Changing the form of the surface of bodies interacting with fields.

Controlling light, light modulation:
- Refraction and reflection of light.
- Electrical and magnetic optical phenomena.
- Photoelasticity.
- The Kerr and Faraday effects.
- The Hahn effect.
- The Franz-Kieldysh effect. This is a change in light absorption when a semiconductor is exposed to an electric field.

Initiation and intensification of chemical changes:
- Ultrasonics.
- Cavitation.
- Ultraviolet, X-ray, radioactive radiation.

· Electrical discharges.
· Shock waves.
· Mycellarian catalysis.
· Quick reactions at high temperatures.
· Nano-sized catalysts

Geometrical Effects:
· Mobius Strip
· Rotating Hyperboloid

Additional Magnetic Effects
- Remanence / hysteresis (memory effect: zero hysteresis to nearly rectangular hysteresis).
- Magnetic particles combined with optically variable pigments.
- Curie point.
- Coercivity. Ability of a magnetic material to withstand an external magnetic field without being demagnetized.
- Barkhausen effect.
- Microencapsulated magnetic particles.
- Magnetic effects in electrochemistry:
    http://electrochem.cwru.edu/ed/encycl/art-m01-magnetic.htm
- Effects of magnetic fields on water:
http://www.lsbu.ac.uk/water/magnetic.html
- Biological effects of magnetic fields:
http://www.biomagnetic.org/magnetic%20effects.html

Controversial Effects:

Controversial phenomena are included in the listing below to maintain open-mindedness. It is left for the reader to make his mind about their validity.

- A list of controversies, including science and technology:

  https://en.wikipedia.org/wiki/Wikipedia:List_of_controversial_issues
- Alfred Wegener's hypothesis of continental drift. Currently the "continental drift" theory is accepted by the mainstream science.
- Tachyons.
- Anomalous phenomenon, (parapsychology)
- Geological phenomenon, (geology)
- Hydrological phenomenon, (hydrology)
- Meteorological phenomenon, (weather)
- Social phenomenon, (sociology)
- Statistical phenomena, (statistics)
- Electronic voice phenomenon (parapsychology)
- Brian Josephson in 2005 declared that parapsychological phenomena might be real . He stated that the eastern mysticism may have relevance to scientific understanding.
- The Pauli effect. The presence of Wolfgang Pauli, apparently caused failure of experiments related to quantum physics, even

if he was in only in transit in a given city. Pauli attributed this effect to the Jungian synchronicity principle. Pauli, in connection with Carl Jung, has developed a theory of "unus mundus", which assumed that there exist a reality, from which everything emerges and to which everything returns. In view of the theory of "unus mundus", the Pauli effect would be a manifestation of Jungian synchronicity, that is a "meaningful coincidence". The theories of Carl Jung were close to the "monopsychism theory", which is the philosophical theory that all humans share the same
eternal consciousness, soul, mind and intellect.
- Bio-effects of magnetism (controversial): http://www.indianetzone.com/2/bio-magnetics_magnetic_effects.htm

[[[]

Selected Additional Physical Effects

- <u>Einstein-de Haas effect</u> - mechanical rotation is induced in a ferromagnetic material (initially at rest), suspended with the aid of a thin string inside a coil, on driving an impulse of electric current through the coil.
- <u>Barnett Effect</u> - the magnetization of a ferromagnetic body when spun on its axis. The magnetization occurs parallel to the axis of spin.

- Bridgman effect (electricity) (electromagnetism)
- Cotton-Mouton effect (magnetism) (optics)
- De Haas-van Alphen effect - oscillation of magnetization as a function of applied magnetic field. The effect can be observed in pure metallic systems at low temperatures and in strong magnetic field, several teslas.
- Garshelis Effect (electric and magnetic fields in matter) (magnetism) (physics)
- Giant magnetoresistive effect **(GMR)** is a magnetoresistance effect, observed in thin film structures composed of alternating ferromagnetic and nonmagnetic metal layers. In the presence of a magnetic field, electrical resistance significantly decreases.
- Hall effect (condensed matter physics) (electric and magnetic fields in matter)
- Inverse Faraday effect (electric and magnetic fields in matter) (optical phenomena)
- Kondo effect (condensed matter physics) (electric and magnetic fields in matter)
- Magneto-optic Kerr effect (electric and magnetic fields in matter)
- Meissner effect (levitation) (magnetism) (superconductivity)
- Proximity effect (electromagnetism) (electrical engineering)
- QMR effect (electric and magnetic fields in matter) (magnetism) (optics)

- Voigt Effect (magnetism) (optics)

Miscellaneous
- List of Emerging Disruptive Technologies:

  http://en.wikipedia.org/wiki/List_of_emerging_technologies

Optical effects
- Asterism, star gems such as star sapphire or star ruby.
- Aura, a phenomenon in which gas or dust surrounding an object luminesces or reflects light from the object.
- Aventurescence, also called the Schiller effect, spangled gems such as aventurine quartz and sunstone.
- The camera obscura
- Caustics - in optics: the envelope of light rays reflected or refracted by a curved surface or object, or the projection of that envelope of rays on another surface.
- Chatoyancy, cat's eye gems such as chrysoberyl cat's eye or aquamarine cat's eye
- Chromatic polarization
- Cathodoluminescence
- Diffraction, the apparent bending and spreading of light waves when they meet an obstruction.
- Dispersion
- Double refraction
- The Double-slit experiment

- Electroluminescence
- Evanescent wave - a nearfield standing wave exhibiting exponential decay with distance. Evanescent waves are always associated with matter, and are most intense within one-third wavelength from any acoustic, optical, or electromagnetic transducer. Optical evanescent waves are commonly found during total internal reflection. The evanescent wave effect has been used to exert optical radiation pressure on small particles in order to trap them for experimentation, or **to cool** them to very low temperatures, and **to illuminate very small objects** such as biological cells for microscopy (as in the total internal reflection fluorescence microscope). The evanescent wave from an optical fiber can be used in a **gas sensor**.
- Fluorescence - a luminescence that is occurring in cold bodies, in which the molecular absorption of a photon triggers the emission of another photon with a longer wavelength.
- Phosphorescence - a process in which energy absorbed by a substance is released relatively slowly in the form of light. Phosphorescent materials glow in the dark after being exposed to light.
- Metamerism - matching of apparent color of objects with different spectral power distributions.
  https://en.wikipedia.org/wiki/Metamerism_(color)
  Metameric matches are quite common, especially in near neutral (grayed or whitish colors) or dark colors. As colors become lighter or more saturated, the range of possible

metameric matches becomes smaller. Making metamerism matches using reflective materials is more difficult than with light-absorbing pigments. The reflective surface colors depend on spectral reflectance curve of the material and the spectral emittance curve of the light source shining. The mineral Alexandrite demonstrates strongly this effect.
- **Pleochroism** - an optical phenomenon in which grains of a rock appear to be different colors when observed at different angles. The effect is caused by the double refraction of light by a colored gem or crystal.
https://en.wikipedia.org/wiki/Pleochroism
- Newton's rings
- Polarized light - related phenomena such as double refraction, or Haidinger's brush
- Rayleigh scattering (Why the sky is blue – atmospheric scattering and absorbance)
- Refraction and the separation of light into colors by a prism.
- Sonoluminescence - the emission of short bursts of light from imploding bubbles in a liquid when excited by sound. Gamma rays could be produced in sonoluminescence experiments (controversial).
- Synchrotron radiation
- Triboluminescence
- The Zeeman effect
- Thomson Scattering

- Total internal reflection
- Twisted light
- The Umov effect - highly reflective objects tend to reflect mostly unpolarized light, and dimly reflective objects tend to reflect polarized light. The law is only valid for large phase angles (angles between the incident light and the reflected light).
- Dichroism - light rays having different polarizations are absorbed by different amounts.
- Dichroic filter - a very accurate color filter used to selectively pass light of a small range of colors while reflecting other colors.
- Goniochromism - the property of certain surfaces to change their colour depending on the angle of view. Examples: are optically variable pigments, pearlescent pigments and some specially woven fabrics or certain shampoos can also exhibit such behaviour.
- Thin-film optics - thin films are used to create optical interference coatings. Applications: optically variable materials, low-emissivity panes of glass for houses and cars, anti-reflective coatings on glasses and for high precision optical filters and mirrors. Another application of thin films is spatial filtering, where an optical device uses the principles of Fourier optics to alter the structure of a beam of coherent light. Spatial filtering is commonly used to "clean up" the output of lasers, removing aberrations in the beam.

APPENDIX D. INTERNET POINTERS AND RESOURCES FOR INVENTORS

PHYSICAL EFFECTS
* Website supporting this Worksheet:
  https://sites.google.com/site/algorithmtriz
* Physical effects with links to descriptions:
  http://en.wikipedia.org/wiki/List_of_effects   (Very Good)
* Electrical Effects and descriptions: Electrical phenomenon (Very Good)
* Optical Effects and descriptions:
  http://en.wikipedia.org/wiki/Optical_phenomenon (Very Good)
* Surface analysis methods:
  http://en.wikipedia.org/wiki/List_of_surface_analysis_methods (Very Good)
* Thermal analysis methods:
  http://en.wikipedia.org/wiki/List_of_thermal_analysis_methods (Very Good)
* Analysis of materials:
  https://en.wikipedia.org/wiki/List_of_materials_analysis_methods

ARTICLES FROM WIKIPEDIA, USEFUL FOR INVENTORS
Physics – glossary:
  https://en.wikipedia.org/wiki/Glossary_of_physics
Outline of physics:
  https://en.wikipedia.org/wiki/Outline_of_physics
Index of physics articles
  https://en.wikipedia.org/wiki/Index_of_physics_articles
Glossary of areas of mathematics
  https://en.wikipedia.org/wiki/Glossary_of_areas_of_mathematics
Glossary of astronomy
  https://en.wikipedia.org/wiki/Glossary_of_astronomy

Glossary of biology
   https://en.wikipedia.org/wiki/Glossary_of_biology
Glossary of calculus
   https://en.wikipedia.org/wiki/Glossary_of_calculus
Glossary of chemistry terms
   https://en.wikipedia.org/wiki/Glossary_of_chemistry_terms
Glossary of engineering
   https://en.wikipedia.org/wiki/Glossary_of_engineering
Glossary of probability and statistics
   https://en.wikipedia.org/wiki/Glossary_of_probability_and_statistics

FIRST CHOICE TRIZ LINKS
   * TRIZ by Glenn Mazur -
   http://www.mazur.net/triz/index.html
   * TRIZ Journal - http://www.triz-journal.com/
   * Inventors' resources, compiled by Ronald J. Riley:
   http://www.rjriley.com/site-index/

BOOKS (out of print and used)
   http://www.bookfinder.com/   - finding out-of-print books and used books
   http://www.alibris.com/   - used books

LIBRARIES
   http://catalog.loc.gov/ - **Library of Congress, Catalog**
   http://www.englib.cornell.edu/ Cornell University - Engineering Library
   http://www.lindahall.org/collections/   Linda Hall Library: the official depository for all ASME publications
   http://www.plos.org/ - Public Library of Science
   https://www.nypl.org/ - NY Public Library

PATENT SEARCHES (multiple sources):
   Google Patents (Very Good, linked patents):
   http://www.google.com/advanced_patent_search

U.S. Patent and Trademark Office:
http://www.uspto.gov/
US Patent Classes - by Title:
http://www.patentec.com/data/class/Classes.htm
Expired US Patents Search:
http://www.uspto.gov/expwd/expform.htm
Canadian Patents Database
http://www.ic.gc.ca/eic/site/cipointernet-internetopic.nsf/eng/h_wr00001.html
World Intellectual Property Office:
https://patentscope.wipo.int/search/en/search.jsf

NATIONAL LABORATORIES AND GOVERNMENT ORGANIZATIONS

http://www.anl.gov   Argonne National
http://www.lbl.gov/   Lawrence Berkeley Laboratory
http://www.lanl.gov/ Los Alamos National Laboratory

WIKIPEDIA LINKS

* Open Access journals
    http://en.wikipedia.org/wiki/Open_access_journal
* Materials Science
    http://en.wikipedia.org/wiki/Materials_Science_and_Engineering
* Publications in Materials Science

http://en.wikipedia.org/wiki/List_of_publications_in_chemistry#Materials_science
* List of scientific journals - Materials science
https://en.wikipedia.org/wiki/List_of_scientific_journals#Materials_science
* List of publications in physics - Materials physics
https://en.wikipedia.org/wiki/List_of_important_publications_in_physics#Materials_physics
* Bio-based materials: http://en.wikipedia.org/wiki/Bio-based_material
* Biomaterial: http://en.wikipedia.org/wiki/Biomaterial

* Liquid crystal:
http://en.wikipedia.org/wiki/Liquid_crystal
* Molecular modelling:
http://en.wikipedia.org/wiki/Molecular_modelling
* Important publications in materials science:

http://en.wikipedia.org/wiki/List_of_publications_in_chemistry#Materials_science

SOCIETIES AND PROFESSIONAL ASSOCIATIONS
http://www.eurekalert.org   American Association for the Advancement of Science
http://aip.org/   American Institute of Physics
http://www.ams.org   American Mathematical Society
http://aps.org/   American Physical Society
http://www.asee.org   American Society for Engineering Education
http://www.acm.org   Association for Computing Machinery
http://www.epri.com/   EPRI: the Electric Power Research Institute
http://www.engc.org.uk/   Engineering Council
http://www.engfnd.org/engfnd/   Engineering Foundation
http://www.iee.org.uk/   IEE - Institution of Electrical Engineers (UK)
http://www.ieee.org/   IEEE
http://www.iop.org   Institute of Physics
http://www.sme.org   Society of Manufacturing Engineers
http://www.spie.org/   SPIE -- **The International Society for Optical Engineering**

MISCELLANEOUS LINKS RELATED TO INVENTIONS
Assistive technology inventions assistance - http://t2rerc.buffalo.edu/
Canadian Innovation Centre - http://www.innovationcentre.ca
Canadian Intellectual Property Office - http://cipo.gc.ca

InventNet - http://www.inventnet.com **(software assisting in writing patents)**

Inventors' Digest - http://www.inventorsdigest.com (Journal)

Inventors' resources, compiled by Ronald J. Riley - http://www.InventorEd.org/inv-reso/

Help for Canadian Inventors ($$$)   https://inventhelp.com

National Institute of Standards and Technology - http://www.nist.gov

United Inventors Association of the USA - http://www.uiausa.com

Finding articles in "popular journals" http://www.findarticles.com/

Quality-Related Site   http://www.quality.org/

INDUSTRY DIRECTORIES
Hovers: http://www.hoovers.com/
Thomas Register: http://www.thomasnet.com/
WebstersOnline: http://www.webstersonline.com/

INFORMATION BY SUBSCRIPTION ($$$)
US Patent Search (QPAT)   http://www.qpat.com/ ($$$)

The Hook, licensing, marketing, and new product development company:
   http://www.thehooktek.com

APPENDIX E. SOFTWARE FOR INVENTORS

There are some TRIZ software products available from several vendors. The software packages are listed below. TRIZ software could be of help in the initial stages of studying the relevant physical effects. Nevertheless, once the physical effects are identified, an in-depth study may require reviewing of the newest and original publications, relevant to the effects, to provide full understanding of the investigated physical phenomena.

TRIZ SOFTWARE (April, 2017)

Vendor: Creax (Belgium)   http://triz.creax.com

Vendor: **Ideation International**
   http://www.ideationtriz.com/

APPENDIX F. BOOKS FOR INVENTORS AND BIBLIOGRAPHY

A. BOOKS FOR INVENTORS

Altshuller, G. (1973). Innovation Algorithm. Worcester, MA.: Technical Innovation Center.
An early book explaining the main concepts of the TRIZ method.

Ball, L. (2002). Breakthrough Inventing With TRIZ. Third Millennium Publishing.

Ball, L. (2002). Hierarchical TRIZ Algorithm. Third Millennium Publishing.

Orloff, M.A. (2006). Inventive Thinking Through TRIZ: A Practical Guide. 2nd ed. Springer Verlag.

Rantanan, K., Domb, E. (2002) Simplified TRIZ. New Problem Solving Applications for Engineers and Manufacturing Professionals. St. Lucie Press, CRC, New York, 2002.
This book explains the basic concepts of the Invention Algorithm: Contradiction, Ideal Final Solution, Ideality of solution and integration of TRIZ with the Six Sigma method.

B. GENERAL BIBLIOGRAPHY

Gawande, Atul (2011). The Checklist Manifesto. How to get Things Right. Picador, 2011.
"Experts need checklists – literally – written guidelines that walk them through the key steps in any complex procedure". Currently, the checklists are in use by the pilots, surgeons,

skyscraper builders and other professionals. Gawande concludes that the checklist provide much needed assurance that all the critical issues are taken care of, and that "progress depends on experts having humility to concede that they need help".

APPENDIX F. FREQUENTLY ASKED QUESTIONS ABOUT TRIZ

QUESTIONS

1. What is Invention Algorithm / TRIZ?
2. Is Invention Algorithm a secret method?
3. Who uses the Invention Algorithm method?
4. Is Invention Algorithm a difficult to learn method?
5. What is the ratio of effects/invested-time in the version of Invention Algorithm
   presented in this text?
6. Are there any books or manuals in English, which would teach the Invention Algorithm/
   TRIZ method?

ANSWERS

1. What is Invention Algorithm / TRIZ?
Invention Algorithm method is a "guided brainstorming" method, which helps in solving technical limitations by using 40 standard principles developed from generalization of creative solutions found in more than 200,000 of selected "breakthrough-type / high quality" patents.

2. Is Invention Algorithm a secret method?

Invention Algorithm / TRIZ is not a secret method, although it has some attributes of a secret method: very little of practical information is published in English (details of the method) and the method is used often in high-tech and defense-related centers. It appears that in the early 1960s, Russians intended to classify this method.

3. Who uses the Invention Algorithm method?

The method is in active use in several high-tech R&D centers in the USA, Russia, South Korea, Germany and Japan. Recently, China started introduction of TRIZ in their companies. Some of the active users in the USA are Ford, Chrysler, General Motors, Honeywell, Rockwell and Xerox. Particularly enthusiastic approach to TRIZ was demonstrated by the industry in South Korea, where the TRIZ method contributed to their "economical miracle" and to spectacular success of such companies as Samsung.

4. Is Invention Algorithm a difficult to learn method?

In the version presented in this text, the method does not require any training. Just follow the step-by-step guide and learn the powerful generalized "40 principles" as you are trying to match them to your problem. In Russia, the TRIZ methods is being taught to high school children.

5. What is the ratio of effects/invested-time in the version of Invention Algorithm presented in this text?

In my opinion, anybody can quickly get 80% of all possible benefits from this method by following the five stages of the "Main Algorithm", supplemented with the standard 40 methods of overcoming technical limitations (usually called "technical contradictions"). Most practical inventors will never need more than the instructions provided in this Workbook.

6. Are there any books or manuals in English, which would teach the Invention Algorithm / TRIZ method?

There are some books explaining the concepts and how to use the TRIZ method. Unfortunately, the most helpful books are still available only in the Russian language. This Workbook is based on publications written in Russian and on own practical experience of the author, who used the method in his professional practice.

www.ingramcontent.com/pod-product-compliance
Lightning Source LLC
Chambersburg PA
CBHW021548200526
45163CB00016B/2981